20 THINGS YOU DIDN'T KNOW ABOUT CENTIPEDES

DID YOU KNOW? BUGS

DOUG BRADLEY

PowerKiDS press

Published in 2027 by The Rosen Publishing Group, Inc.
2544 Clinton Street, Buffalo, NY 14224

Portions of this work were originally authored by Matthew Elkin and published as *20 Fun Facts About Centipedes*. All new material in this edition was authored by Doug Bradley.

Editor: Greg Roza
Book Design: Michael Flynn

Photo Credits: Cover reptiles4all/Shutterstock.com; series background freshcare/Shutterstock.com; p. 4 phichak/Shutterstock.com; p. 5 Lauren Suryanata/Shutterstock.com; p. 6 kooanan007/Shutterstock.com; p. 7 Maximillian cabinet/Shutterstock.com; p. 8 Audrey Snider-Bell/Shutterstock.com; p. 9 Sytilin Pavel/Shutterstock.com; pp. 10, 25 I Wayan Sumatika/Shutterstock.com; p. 11 Abimanyuabii/Shutterstock.com; p. 12 Igor Kovalchuk/Shutterstock.com; p. 13 SherSS/Shutterstock.com; p. 14 prasom boonpong/Shutterstock.com; p. 15 Henrik Larsson/Shutterstock.com; p. 16 Nick Greaves/Shutterstock.com; p. 17 Azmil Umry/Shutterstock.com; p. 18 Alfredo Peralta Garcia/Shutterstock.com; p. 19 chinahbzyg/Shutterstock.com; p. 20 Mike Kovalenko/Shutterstock.com; p. 21 WITSANU PORNSUKNIMITKUL/Shutterstock.com; p. 22 gan chaonan/Shutterstock.com; p. 23 kooanan007/Shutterstock.com; p. 24 MATATABINOTABI/Shutterstock.com; p. 26 Maximillian cabinet/Shutterstock.com; p. 29 Colombe Photographie/Shutterstock.com.

Library of Congress Cataloging-in-Publication Data

Names: Bradley, Doug, 1971- author
Title: 20 things you didn't know about centipedes / Doug Bradley.
Other titles: Twenty things you didn't know about centipedes
Description: Buffalo : PowerKids Press, [2027] | Series: Did you know? Bugs | Includes bibliographical references and index.
Identifiers: LCCN 2025048926 (print) | LCCN 2025048927 (ebook) | ISBN 9781499457377 library binding | ISBN 9781499457360 paperback | ISBN 9781499457384 ebook
Subjects: LCSH: Centipedes–Miscellanea | Centipedes–Juvenile literature
Classification: LCC QL449.5 .B73 2027 (print) | LCC QL449.5 (ebook)
LC record available at https://lccn.loc.gov/2025048926
LC ebook record available at https://lccn.loc.gov/2025048927

Printed in China

CONTENTS

LOOK AT ALL THE LEGS!

Centipedes are arthropods, which are animals that have a jointed **exoskeleton**, jointed legs, and no backbone. They're amazing creatures that are hard to find and often misunderstood.

Centipedes live all over the world. Most are nocturnal, which means they're active at night. This makes them hard to find during the day, when they often try to hide under leaves and rocks and in other dark, damp places. Centipedes may look creepy, but they help keep your home free of other creepy creatures.

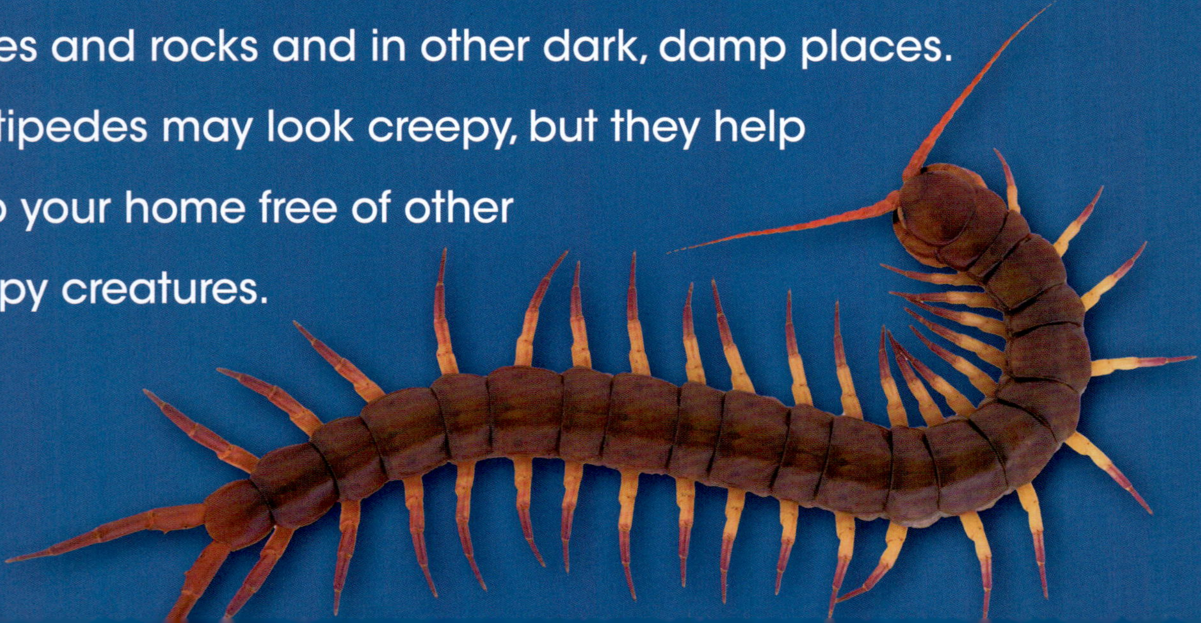

So far, about 3,000 species, or kinds, of centipedes have been identified. Some scientists think there may be far more yet to discover.

ODD LEGS

FACT 1

CENTIPEDES CAN HAVE MANY DIFFERENT NUMBERS OF LEGS.

Centipedes are easy to spot in the wild because of their long bodies and numerous legs. Most people think that centipedes have 100 legs, but they usually don't. They can have as few as 30 legs or more than 300 legs!

The word "centipede" comes from the Latin words for "hundred" and "foot."

CENTIPEDES ALWAYS HAVE AN ODD NUMBER OF LEG PAIRS.

FACT 2

Centipede bodies are made up of many **segments**. Each segment has two legs. Centipedes always have an odd number of leg pairs. Depending on the species, adult centipedes can have as few as 15 leg pairs, like the house centipede, or as many as 191.

FACT 3 NOT ALL OF A CENTIPEDE'S LEGS ARE JUST LEGS.

A centipede's first pair of legs below the head is a set of **venomous** pincers. Centipedes use these legs like jaws to "bite" **predators** and catch **prey**. The last pair of legs are usually longer than the rest of the legs.

The back legs are longer to help the centipede move faster.

CENTIPEDE ANATOMY

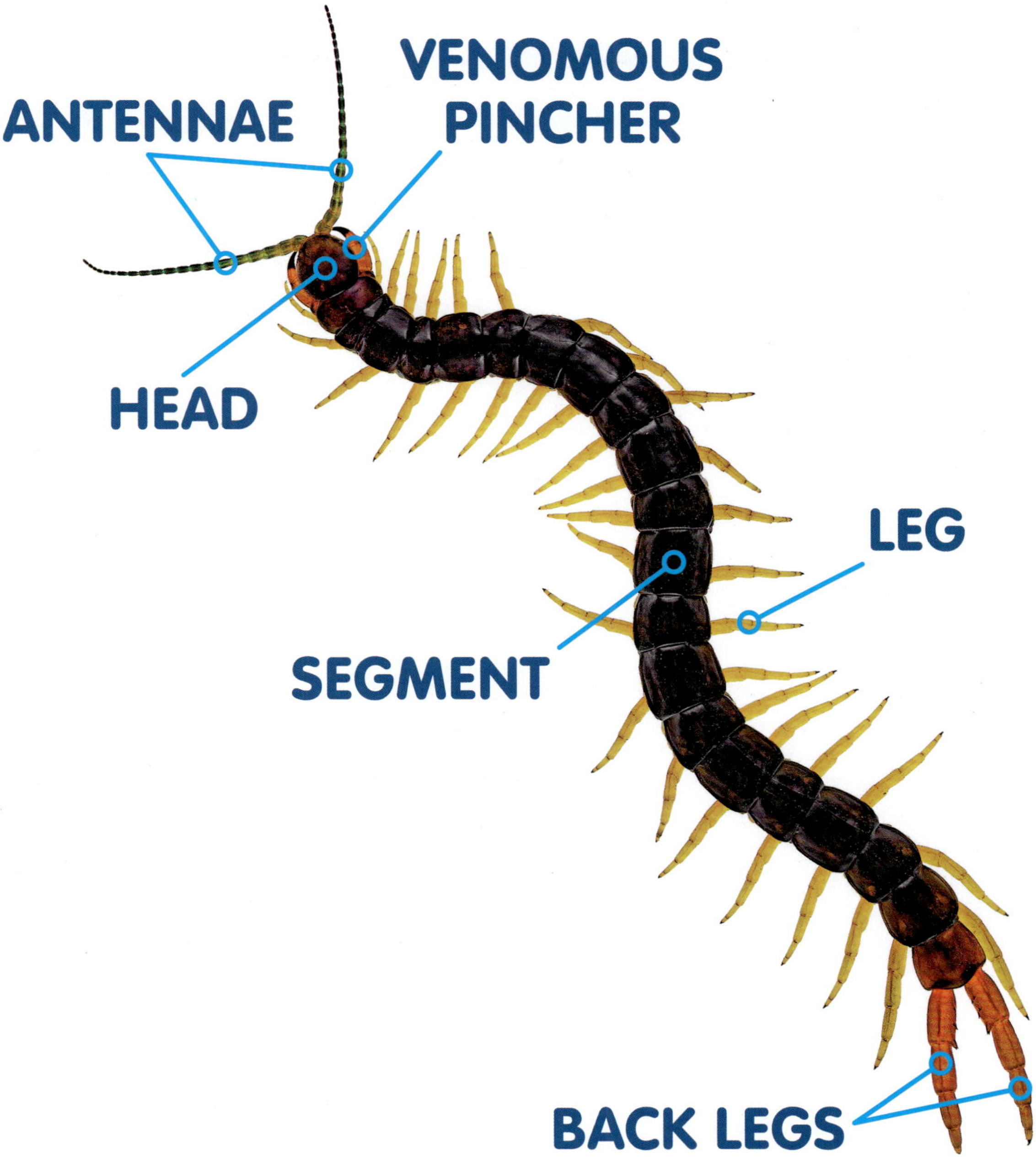

CENTIPEDE BODIES

FACT 4

THE BIGGEST CENTIPEDE CAN GROW UP TO 1 FOOT (30 CM) LONG.

Most centipedes are tiny. The smallest centipedes are just a few millimeters long. However, the Amazonian giant centipede can grow to 12 inches (30 cm) long. That's longer than an adult's hand!

Some people keep giant centipedes as pets.

CENTIPEDES DON'T BREATHE THROUGH THEIR MOUTH.

FACT 5

Centipedes breathe through holes in their exoskeleton called spiracles. Some species have spiracles on each segment, while others don't. Centipedes can't close their spiracles. This would cause their bodies to dry out.

FACT 6

CENTIPEDES ARE MORE LIKE LOBSTERS AND CRABS THAN INSECTS.

Centipedes aren't insects, even though we call them bugs. They're arthropods, like lobsters and crabs. Arthropods have segmented bodies, a hard exoskeleton, jointed legs, and many pairs of limbs. Arthropod bodies have symmetry, which means they look the same on both sides.

Centipedes are a type of arthropod called myriapods (MIH-ree-uh-pahds). "Myriad" means "many," and "pod" means "foot."

THAT'S LIFE!

FEMALE CENTIPEDES LAY BETWEEN 15 AND 60 EGGS.

FACT 7

Female centipedes can have up to 60 babies at one time, sometimes more, in soil or rotting wood. The eggs can take from one to several months to hatch. Once a baby centipede is born, it takes several years for it to become an adult.

FACT 8

SOMETIMES, MOTHER CENTIPEDES EAT THE EGGS THEY LAY!

Female centipedes are known to stay close to their nests and keep their eggs safe. They lick the eggs to keep them clean. However, if something bothers the nest before the eggs hatch, the female may leave the eggs—or eat them!

Female centipedes often curl their body around their eggs and newly hatched young to keep them safe.

SOMETIMES, BABY CENTIPEDES EAT THEIR OWN MOTHER!

FACT 9

Some species of centipedes are matriphagic (ma-truh-FAY-jihk). This means the baby centipedes actually eat their mother soon after they hatch! The babies use the **nutrients** from their mother's adult body to help them begin to grow.

FACT 10

AS MANY CENTIPEDES GROW LONGER, THEY GROW NEW SETS OF LEGS AS WELL.

Centipedes may have no more than four pairs of legs when they are young. As they grow, they shed their skin, or molt, revealing a bigger body with new legs each time. Centipedes can go through as many as 10 molts during their lifetime.

The blue-headed centipede, shown here, is common in Thailand. It's very fast and is known to bite.

MOST CENTIPEDES HATE THE SUN.

FACT 11

Centipedes stay hidden most of the day. That's why we don't always see them. The sun's heat causes their bodies to lose water quickly, causing them to dry out and die. They prefer to hunt at night, when they're safe in the cool, moist air.

During the day, centipedes may hide on the underside of a leaf to stay cool and hidden.

FACT 12 MANY PEOPLE SHARE THEIR HOMES WITH CENTIPEDES AND DON'T EVEN KNOW IT.

Centipedes usually live outdoors. They're often found under rocks and in deep cracks. However, they also like damp areas inside your home. House-dwelling centipedes can hide in dark basements, bathrooms, and closets—and can live there their entire life.

House centipedes, like this one, live in homes to stay away from the hot summers and cold winters.

SENSING THE WORLD

CENTIPEDES CAN'T SEE VERY WELL.

FACT 13

Some centipedes have compound eyes. Most have groups of simple eyes on each side of their head. These eyes are called ocelli (oh-SEH-lee). Ocelli identify light and dark, but that's about it. They're not very strong. Some centipedes don't have eyes at all!

This centipede has compound eyes. Each is made up of thousands of lenses!

Centipedes use their antennae to locate food.

FACT 14 CENTIPEDES CAN'T SEE WELL, BUT THEIR OTHER SENSES MAKE UP FOR IT.

Centipedes use their senses of smell and touch to find prey. As nocturnal creatures, these senses help them to stay safe and well fed. Antennae at both ends of their body help them find their way around.

FEARSOME PREDATOR

CENTIPEDES ARE BIG EATERS.

FACT 15

Centipedes are **carnivores**. They feed mostly on smaller arthropods. Some larger centipedes have been known to attack and feed on frogs, toads, small mammals, snakes, birds, and bats.

This centipede has caught a tasty bullfrog for dinner.

FACT 16

CENTIPEDES ARE THE ONLY ANIMALS ARMED WITH FORCIPULES.

A centipede's front legs are also venomous pincers. They're called forcipules (FOHR-suh-pyulz). Centipedes use their forcipules to **inject** a powerful venom into their prey. The larger the centipede, the stronger and more deadly the venom is.

The centipede's venomous pincers make them different from all other creatures in the animal kingdom.

Centipede bites cause redness, swelling, and pain.

CENTIPEDE VENOM CAN BE DANGEROUS FOR SOME PEOPLE.

FACT 17

Centipede venom is much more dangerous for small animals. Centipede bites can feel like painful bee stings, but are rarely deadly. People who are **allergic** to bee stings, however, can also have bad reactions to centipede venom.

DINING ON CENTIPEDES

FACT 18

LOTS OF ANIMALS EAT CENTIPEDES, INCLUDING PEOPLE!

Centipedes are prey for many larger predators. These include birds, toads, rats, spiders, and other centipedes. Many centipedes are stepped on by creatures that don't see them. In some parts of the world, even people enjoy eating centipedes!

Centipedes and other insects are common foods in parts of the world. Would you try one?

Scorpions and centipedes are both venoumous. Which do you think would win this battle?

CENTIPEDES CAN ESCAPE PREDATORS BY LOSING LEGS!

FACT 19

Some centipedes have bright colors that warn predators they're no good to eat. Others use **camouflage** to hide from predators. Many centipedes can drop off legs when predators grab onto them. This allows them to run away. The legs grow back!

EVEN MORE LEGS!

FACT 20

MILLIPEDES ARE CLOSELY RELATED TO CENTIPEDES, BUT THEY HAVE MORE LEGS.

Centipedes are sometimes confused with their many-legged relatives, millipedes. Unlike their carnivorous cousins, millipedes are slow-moving, plant-eating myriapods. They have two pairs of legs on most body segments instead of the single pair that centipedes have.

MILLIPEDE

Centipedes and millipedes are often confused with one another, but they're very different.

WHAT'S THE DIFFERENCE?

CENTIPEDES

- ONE PAIR OF LEGS PER SEGMENT
- USE FORCIPULES TO INJECT VENOM
- FLAT BODIES
- MEAT EATERS
- FAST RUNNERS

BOTH

- ARTHROPOD
- LIVE IN DAMP LEAVES AND BASEMENTS

MILLIPEDES

- SLOW MOVING
- PLANT EATERS
- ROUNDED BODIES
- COIL UP TO GUARD THEMSELVES
- TWO PAIRS OF LEGS PER SEGMENT

Centipedes and millipedes have several key differences. However, they're both arthropods.

RUNNING WITH THE CENTIPEDES!

Centipedes have been around for over 400 million years! Their ancient relatives were some of the first creatures to walk on land. They've lived so long because of their abilities to crawl into deep, dark hiding places and guard themselves with their powerful venom and sharp forcipules.

You may not see them a lot, but there's bound to be some creeping under rocks and in basements near you. If you see a centipede, try to count how many legs it has before it runs away!

Centipedes look creepy! But house centipedes eat other pests, such as cockroaches and silverfish.

GLOSSARY

allergic: To be sensitive to normally harmless things in the surroundings, such as dust, pollen, or mold.

camouflage: Colors or shapes in animals that allow them to blend with their surroundings.

carnivore: An animal that only eats other animals.

exoskeleton: The hard outer covering of an animal's body.

inject: To use sharp teeth or claws to force venom into an animal's body.

nutrient: Something a living thing needs to grow and stay alive.

predator: An animal that hunts other animals for food.

prey: An animal that is hunted by other animals for food.

segment: A part of an insect's body.

venomous: Able to produce a liquid called venom that is harmful to other animals.

FOR MORE INFORMATION

BOOKS

Aicher, Hope. *Centipedes and Millipedes!* Independently published. My Incredible World, 2025.

Becker, Trudy. *Millipedes*. Mendota Heights, MN: North Star Editions, 2025.

WEBSITES

Centipede
animalfact.com/centipede
Learn more about this many-legged arthropod at this detailed website.

Centipede Facts
www.activewild.com/centipede
This resource has fascinating centipede facts, close-up photographs, and links to related topics.

Publisher's note to educators and parents: Our editors have carefully reviewed these websites to ensure that they are suitable for students. Many websites change frequently, however, and we cannot guarantee that a site's future contents will continue to meet our high standards of quality and educational value. Be advised that students should be closely supervised whenever they access the internet.

INDEX